1 0

ers, 1999

is book is available from the British Library

reat Britain by The Bath Press

Cover image:
Lightbulb © PhotoDisc, Inc

HarperCollins Publishers
PO Box, Glasgow G4 0NF

First published 1999

Reprint 10 9 8 7 6 5 4 3 2

© HarperCollins Publish

ISBN 0 00 472326 0

A catalogue record for th

Printed and bound in G

Contents

Illustration Credits

Introduction

A book such as this, highlighting the apparently inventive 'nature' of the Scottish nation, raises a simple question: 'Why the Scots?' In a remarkable range of fields the Scots contribution has been, and continues to be, crucial. So much so, that this seems to be the result of a national characteristic.

Scotland itself is not the harshest place to live, although to early settlers it may often have seemed that way. Stuck on the outskirts of the known world, Picts and Scots had to devise methods of sheltering and scraping a means of survival that may be the root of this phenomenon.

Perhaps the weather has had more of a hand in this than we might normally acknowledge. For, once shelter has been dealt with, continual rain becomes more of a pest than a hardship. It is not too hard to imagine our forebears, with no electronic distraction, looking for a way to while away the rainy hours indoors.

As the nation grew in the shadow of a larger neighbour, so too did a defiant streak and, sometimes, over-confidence: the disastrous failure of the ambitious Darien scheme, Scotland's attempt to colonise Central America, had repercussions still felt today, for it was a strong factor in persuading the Scottish gentry of the benefits of a United Kingdom. The other side of the 'small nation' coin is a tendency to belittle the achievements of fellow Scots, known as the 'I kent his faither' mentality. It should be no surpise that many inventors left their homeland before making the breakthrough to success.

The country's religious background, too, has been a force in moulding the characteristics of the people. The Calvinist work ethic has reaped rewards for many, with the Scots taking to heart, more than most, the saying, 'The Devil will find work for idle hands to do'.

There is no truth in the claim that copper wire was invented by two Scotsmen fighting over a penny they saw on the pavement. However, as youwill see in the following pages, the Scots have learned to make the most of available resources.

Alistair Fyfe

SCOTS have always looked beyond the boundaries of their small country for information and inspiration. Some have stayed put and seen their ideas travel, while others have wandered the globe putting their inventiveness to purpose.

Observing and Measuring

James Gregory (1638-75), Aberdeenshire-born professor of Mathematics at St Andrews and Edinburgh, invented the reflecting telescope in 1661, which was developed by Sir Isaac Newton. Newton's theories of gravitation were in turn promoted in print by **David Gregory** (1659-1708), James' nephew.

James Ferguson (1710-76) was an amateur astronomer for many years before landing a professional post. He began life as a farm labourer's son and used his sheep-tending duties as an opportunity to watch the stars. His diverse other jobs along the way included servant and even portrait painter. He devised many mechanical contraptions to illustrate the workings of the solar system, including a tide-dial, an eclipsarion (which illustrated a solar eclipse) and various astronomical clocks.

Robert Blair (1748-1828) was a minister's son, born at Garvald in East Lothian. Originally a surgeon, he became the first Regius Professor of Astronomy at Edinburgh University in 1785. His research concentrated on the effectiveness of reflecting telescopes, and experiments with fluid lenses led him to create widely acclaimed telescopes of several inches.

The urge to identify, classify and measure the world around them has spurred many inventors. Fife-born Sir **John Leslie** (1766-1832) travelled extensively in Europe and America. His inventions on his return were numerous and usually concerned with measurement. They include a differential thermometer (for measuring difference of temperature), a hygrometer (measuring humidity), a photometer (measuring luminous intensity), the atmometer (measuring the rate of evaporation from a moist surface) and the aethrioscope (measuring minute variations in temperature due to the condition of the sky). He succeeded in creating artificial ice in 1810.

Robert S Newall (1812-89) combined engineering with interests in astronomy, but it was the former field he was more innovative. He patented a new type of wire rope which he manufactured at Gateshead. Then, considering the uses of cables for communications, he developed improved cables for submarine use, laying such cables across the world.

The New World

The Scots contribution to the maritime world ranges from the building of the great ships, to the smallest aid to sailors.

James Paterson (1770-1806) of .Musselburgh developed the machine process for manufacturing fishing nets.

For all Scotland's remoteness in Europe, its location at the western edge of the old world was to be one of its great assets. It could even be argued that America was the most important discovery in Scotland's history. Clan Sinclair claim that Prince **Henry Sinclair** (b1345) travelled via Nova Scotia to what is now Massachusetts in 1398, nearly a century before Columbus 'discovered' the continent. In time, with the establishment of the Greenwich meridian, Britain became the centre of the world and Glasgow was a jumping-off point for the New World, leading to the city's supremacy in shipbuilding.

Once in America, many Scots have thrived on the opportunities afforded by the New World. **Andrew**

Andrew Carnegie : a New World Scot

Carnegie's name is synonymous with wealth, and a great example of a Scot who remembered his roots. Other names are not widely identified as Scottish.

David Dunbar Buick (1854-1929) is one less well known as a Scot. Taken to America from Arbroath when still an infant, Buick first worked as a farm labourer, then wholesale plumber, before experiments with the new gasoline motors grew to the establishment of the Buick Manufacturing Company, which made car engines. The first Buick car was built in 1903 and the company was taken over by General Motors in 1908.

John Muir (1834-1914) was born at Dunbar. He attributed his interest in the wild places of America to his boyhood in the Scottish countryside. He fathered the environmental movement, persuading the US Government to establish the first National Park at Yosemite.

Sir **Frank Fraser Darling** (1903-79) was born in Edinburgh and educated at the Midland Agricultural College and the Institute of Animal Genetics of Edinburgh University. His interest in animal ecology took him on world-wide travels arguing the case for the protection of endangered species.

Adam Ferguson (1723-1816) was Professor of Moral Philosophy at the University of Edinburgh. He began the study of humans in groups, the subject now known as Sociology.

John Muir

Inventive Links

The continuing tradition of inventiveness perpetuated itself as one invention led to another, with education and training encouraging self-improvement and creativity. This background established the chains of events which led to series of inventions, and encouraged diversity. An example of this is Robert Napier.

Robert Napier (1791-1876) left Robert Stevenson's (see p. 20) employ to set up his own foundry in Glasgow, making pipes for water and gas. After a commission to make components for marine engines he began building those engines himself, eventually moving on to build ships themselves. Ships for the East India Company were followed by Navy contracts, and Napier was established as a ship-builder of renown, one of those who made 'Clydebuilt' a byword for quality in shipping.

Robert Napier: start of a chain of inventiveness

His partnership with **George Burns** (1795-1890) and Samuel Cunard created the British & North American Royal Mail Steam Packet, later called the Cunard Line. Burns and his brother James were among Glasgow's first home-grown shipping magnates.

William McNaught (1813-81) of Paisley was apprenticed to Robert Napier. He later worked with his father's firm of engine component manufacturers, helping supply the growing manufacturing industry which increasingly used steam-driven machinery. In response to a demand for a more powerful engine he devised an extra cylinder, operating at a different pressure from the first. The success of this invention led to the nickname 'McNaughting' to describe the addition of a second cylinder.

David Kirkaldy (1820-97) was an apprentice at Robert Napier's Vulcan Foundry in Glasgow shortly after McNaught left the company, training as a draughtsman. Having established methods for testing strengths of iron and steel at Napier's, he moved to London, establishing the first independent materials-testing laboratory. The machinery in the laboratory was built to Kirkaldy's own designs. The draughtsmanship learned at Glasgow was carried out to such a fine art that his engineering drawings were exhibited at the Royal Academy and the Louvre.

John Elder (1824-69), born in Glasgow, was the son of a manager at Napier's shipyard and served his own apprenticeship there. After a period in charge of the drawing office at Napier's, Elder left to form his own company in partnership with Charles Randolph. Randolph, Elder & Company later became better known as the Fairfield Shipbuilding Company. Their marine compound steam engine was so efficient by comparison with others that it established the yard at the forefront of marine engineers. The yard at Govan, now owned by Kvaerner, is one of the few remaining Clyde shipyards, still producing leading-edge shipping technology.

Alexander Carnegie Kirk (1830-92), born in Forfarshire and another Napier's apprentice, produced the triple-expansion engines that eventually superseded Elder's compound engine. The great advance in technology of the day, the engine even featured in a poem by William McGonagle. Kirk spent a period working for James Young at Bathgate, making improvements to the shale-oil production methods, before returning to marine engineering.

James Napier, Robert Napier's son, followed his father's trade after graduating. He pioneered the used of iron in shipbuilding, when the government was looking for stronger warships. He was consulted when crafts were required for the shallow waters of the Indian sub-continent, producing a steamer with a draft of only one foot, and invented a method of adjusting compass readings in iron ships He turned his attention to domestic appliances, inventing the Napier Glass Coffee Apparatus, the percolator now common, where water in the bottom section is heated to boiling, the pressure forcing it up through the coffee into the top of the pot. His name is no longer associated with this invention, for he did not apply for a patent.

Kelvin

The career of **William Thomson** (1824-1907), 1st Baron Kelvin of Largs, has so many aspects that he cannot be categorised. This Irish-born intellectual brought his inventiveness to his new home when he was six, after the death of his Scottish mother, and always considered himself Scottish.

The young Thomson's arrival in Glasgow followed the appointment of his father as Professor of Mathematics at the University. The prodigious child had never been to school, for his father taught the family. He entered the University at the age of 10, and studied at Cambridge University and at Paris. He was appointed Professor of Mathematics and Natural Philosophy at the Glasgow when still only 22 and turned his mind to applied and pure physics. By the age of 26 he had published fifty original papers, mostly mathematical, several in French. He was determined to measure and identify properties of matter, saying it was

William Thomson, Baron Kelvin

Kelvin's historic last lecture at the University of Glasgow in 1899

necessary to fix on 'something absolutely as the measure of reckoning'. He proposed the absolute temperature scale in 1848, now known as the Kelvin scale. Further studies laid the basis of the study of thermodynamics, but Kelvin generously gave credit to his theories to the other scientists who had helped with the initial ideas. He was not satisfied with the theoretical; his work had to have practical applications. From his theories regarding heat transfer and the dissipation of energy, he suggested the principle of refrigeration. His findings on electrical oscillation formed the basis of wireless telegraphy. These were developed by others, but Kelvin took a personal hand in the development of long-distance communications. He was chief consultant on the laying of the first underwater Atlantic cable, and his wealth was founded on his patent for a mirror galvanometer which improved the speed of telegraphic transmissions. It has been speculated that the idea for the mirror came from light reflected from the monocle which the short-sighted Kelvin wore on a ribbon.

KELVIN'S INVENTIONS: A SHORT LIST

- a device for the absolute measurement of currents
- the mirror galvanometer
- the electric strain gauge
- the siphon recorder
- the Kelvin ampere balance
- the electrostatic voltmeter
- the doctrine of available energy
- the theory of electrical oscillations
- the principle of the heat pump
- the development of the wave theory of light
- the development of molecular dynamics
- the development of the second law of thermo-dynamics

His inventions included several for mariners, namely a tide predictor and improved ships' compasses. He also invented electrical instruments which were made by his own company, Kelvin & White, a partnership with a local optical instrument maker. His home in Glasgow was the first to be lit by electricity.

Married twice, Kelvin settled in Largs, Ayrshire in later life, and it was from this home that he adopted his title when raised to the peerage in 1892, of Baron Kelvin of Largs. Altogether some 661 scientific papers are attributed to Kelvin.

James Thomson (1822-92), Lord Kelvin's elder brother, was an authority on hydraulics. Professor of Engineering at Belfast University and then at Glasgow, he invented a turbine and wrote papers on currents and winds.

$\int$ COTLAND is one of the most remote parts of Europe, with a climate and terrain both equally inhospitable. The Scots' inventiveness has often turned to ways of overcoming these features, to help people move around or to leave the place altogether. The first stage was to improve the communication lines, then the methods of moving along those routes.

Roads

John Loudon McAdam (1756-1836) was born in Ayr. In his teens he travelled to New York, earning his fortune before returning to Scotland in 1783. He took over the British Tar Company, and on his new Ayrshire estate he experimented with methods of road building which he later applied to road construction for the Bristol Turnpike Trust. His methods of improving drainage and creating a more permanent road surface were dubbed 'macadamisation' and his advice was much sought.

However his travel and experimentation were often at his own expense and having spent his fortune he sought help from Parliament in recognition of his work, eventually being awarded £2000, in 1820. In 1827 he was made Surveyor-General of metropolitan roads. McAdam's contribution was quite literally the groundwork for future Scots' inventions.

John Loudon McAdam

Bicycles

Kirkpatrick Macmillan (1813-78) was born near Thornhill in Dumfriesshire. After working as a farm labourer and a coach-man he took up the trade of his father: blacksmith. He made himself a copy of a hobby-horse, a bicycle-like contraption which the rider propelled by pushing his feet on the ground. Unsatisfied with this means of propulsion, Macmillan experimented with pedals and cranks, put pedals on a tricycle as early as 1834, and eventually devised the first bicycle in 1840. He rode his new device as far afield as Glasgow, a distance of 70 miles, in two days. Macmillan never patented his invention, and it was widely copied, that invention was even being credited to a Lesmahagow man, Gavin Dalzell.

Thornhill Smithy, birthplace of the bicycle

Pneumatic Tyres

Robert William Thomson (1822-73) is not recognised as the inventor of the pneumatic tyre, but it was he who patented the principle in 1845. Born in Stonehaven, Thomson applied his inventive mind to various fields – perhaps a reason why his vulcanised rubber pneumatic tyre was not developed at the time. The Thomson tyre was successfully tested in the 1840s but thought too expensive, eventually being forgotten.

Thomson left his native land for the sugar plantations of Java, where his work as an engineer led him to improve on the production machinery, and to design the first mobile steam crane. In 1867 he patented a steam traction engine. The 'Thomson Steamer' was produced commercially from 1872 and many were exported around the world in the next two decades.

John Boyd Dunlop (1840-1921) resurrected Thomson's principle four decades later. Dunlop was born in Dreghorn, Ayrshire. He became a veterinary surgeon, working in Edinburgh and then Belfast. In 1887 his fatherly concern led him to replace the solid rubber tyres of his child's tri-cycle with inflated rubber hoses. The success of this

The first pneumatic tyre factory i the world was started here in 1889. t make tyres under John Boyd Dunlop patent of the 7th December, 1888.

early experiment led to the creation in 1889 of the business which became the Dunlop Rubber Company Ltd, making his name synonymous with rubber products.

Sir Robert McAlpine (1847-1934) earned the nickname 'Concrete Bob' as a pioneer the use of concrete and of labour-saving machinery in the construction industry.

OTHER SCOTTISH ROAD TRANSPORT FIRSTS

- **Henry, Lord Brougham** designed the Brougham one-horse carriage in 1838
- Postman **Andrew Lawson** invented the steam tricycle, nicknamed the Craigievar Express, in 1895
- **William Murdoch** (1754-1839) invented a steam car
- **John Yule** of Glasgow built the first motor lorry in 1870
- **Sir Keith Elphinstone** (1864-1944) invented the speedometer
- a Scottish-inspired Act of Parliament in 1772 first determined the rule of driving on the left

$\mathcal{R}$OADS alone would not take the traveller on the most direct routes from A to B. The coastline of Scotland is longer than the eastern seaboard of the USA, but few people want to travel via the scenic route. The development of methods of crossing rivers, firths, chasms and gorges has tested the engineering genius of many Scots.

Bridges

Sir **Samuel Brown** (1776-1852) was born in London of Scottish parents, and although he had a distinguished naval career his achievements in bridge design are of more renown. His main patents were registered in 1816 and 1817, covering components of suspension bridges. He built bridges throughout Britain, but his Union Bridge over the River Tweed is seen as the finest. With a span of over 440 feet, the bridge was the first suspension bridge in the country to carry the heavier loads presented by carriages and loaded carts. Built in 1820, the bridge still carries limited road traffic.

Sir **William Fairbairn** (1789-1874), born in Kelso, moved south and when in his teens was an apprentice engine-wright in North Shields. It was here he befriended the young George Stephenson, later to invent the locomotive. Fairbairn's career saw him making cotton machinery for Manchester mills before experiments steam engines and iron boat building led him to establish a yard at Millwall which was a prolific producer in the 1830s and '40s before turning his attention to the design of bridge components, inventing the rectangular tube used by Robert Stephenson (son of George) for his railway bridges at Menai and Conway. Tubular steel is stronger than solid steel, and so became integral to many hundreds of bridges.

Perhaps Scotland's most famous bridge builder was Sir **William Arrol** (1838-1913), born in Houston, Renfrewshire. Almost inevitably at that place and time his first job was in a cotton mill, but at 14 he became an apprentice blacksmith, also study-

Civil Engineering

Arrol's masterpiece: the Forth Rail Bridge

ing at night school. In 1868 he established his own engineering business, Dalmarnock Iron Works, in Glasgow. After the disastrous collapse of the Tay Bridge it was Arrol who won the contract for the replacement, but it was his Forth Bridge which became one of the wonders of the modern world. Arrol's also built Tower Bridge in London in a period which saw the civil engineer become essential to the development of industry.

Railways

Many of Arrol's bridges were built at a time when the railways were the burgeoning industry, revolutionising transport of people and goods. Of course Watt's steam engine (see p. 23) was central to the success of the train, and across the world Scots pioneered its use.

Glasgow-born Sir **John MacDonald** (1815-1891) was Canada's first Prime Minister and a central figure in the creation of the Canadian Pacific Railway.

Roads

If Arrol is the most famous Scottish bridge engineer it is only because **Thomas Telford** (1757-1834) is more usually known as the 'Colossus of Roads', although his output included many bridges, harbours, canals and other engineering projects. Telford was born in 1757 in Westerkirk, Langholm, the son of a shepherd. Apprenticed to a stonemason, he went to Edinburgh in 1780 and to London two years later. Following periods at the Portsmouth dockyard and as surveyor of public works for Shropshire, he was commissioned to report to the government on public works for Scotland. At the time, in 1801, Scotland's rural roads were still in the main the old system of military roads built in the wake of the 1745 Jacobite rebellion. The years follow-

Thomas Telford

ing saw more than 1000 miles of roads constructed under Telford's supervision, including over 1200 bridges. Outside Scotland, he also designed the bridge over the Severn at Montford (1790-92), the Menai Suspension Bridge (1819-26) and London's St Katherine Docks (1824-28). Telford's career spanned many engineering works including the more important pre-railway routes for goods transport – the canals.

Canals

The waterways of Britain have been called the power lines of the early industrial society, for along these routes travelled the coal which powered the early factories. Canals may not have been invented by Scots but their development in Britain inspired the huge growth of the industry.

Thomas Telford was engineer and architect of the Ellesmere Canal. Sir Walter Scott described Telford's aqueduct at Pont

Cysylite as 'the most impressive work of art I have ever seen.' He went on to build the Caledonian Canal, the Gloucester Canal and, in Sweden, the Gotha Canal.

Docks

The importance of water transport has waned recently, but in a country of firths and sea lochs, people had taken to the water from the earliest times. The need to cross large expanses of water was met by a series of inventions in the fields of boat building and navigation.

John Rennie

John Rennie (1761-1821), born at Phantassie, East Lothian, built docks from northern Scotland to south-western England, including the London and East India Docks, and Southwark, Waterloo and London Bridges (the London Bridge he began was finished by his son, also John, and was later dismantled and re-built in Arizona, USA). His eldest son George (1791-1866) built the first screw vessel for the Royal navy, the *Dwarf*.

Thomas Morton (1781-1832) invented the patent slip for docking vessels, a cheap alternative to the dry dock, in 1819.

James Bremner (1784-1856) from Keiss, Wick, was apprentice at a Greenock shipbuilder and went to America before coming home to take over a shipyard. Besides building ships and salvaging sunken or stranded ships he built harbours and piers in the far north, notably at his native Keiss and at Lossiemouth. He invented a new type of crane and designed a new pile driver.

Lighthouses

Sir **David Brewster** (1781-1868), born in Jedburgh, specialised in the field of optics, and his contribution to shipping was in the design of lighthouses, where he persuaded the authorities to adopt dioptric lenses. (See also p. 57.)

But the greater contribution to lighthouse design and function was by Glasgow-born **Robert Stevenson** (1772-1850).

Stevenson was still young when his father died, but when his mother remarried it was his stepfather Thomas Smith's career

Robert Stevenson

he followed. Stevenson succeeded Smith as first engineer of the Northern Lighthouse Board in 1779. In 47 years he built 23 lighthouses in Scotland, including the Bell Rock off Arbroath, and invented a system of flashing or intermittent lights. His family took to the profession: three sons, two grandsons and a great-grandson were lighthouse engineers, continuing the line as engineer to the Northern Lighthouse Board until 1938. (One notable exception to the family career

was grandson Robert Louis Stevenson.) It was the earnest hope of these lighthouse designers that their innovations would save as many lives as possible. In a spirit of altruism they deliberately did not patent their inventions, hoping that they would be widely copied.

Thomas Drummond (1797-1840), born in Edinburgh, entered the Royal Engineers in 1815, and joined the Ordnance Survey in 1820. He invented the 'Drummond Light', of benefit to the

Civil Engineering

OS, and ultimately to good mariners. The invention enabled observation of far distant points and Drummond adapted its use for lighthouses. As Secretary of State for Ireland he gained popularity among the people with his attitude to absentee landlords, typified by the phrase he coined: 'property has its duties as well as its rights'.

As the inventors of one age passed the baton to the next generation, there were plenty willing new spirits to take up the challenge. Just as Stevenson's family were establishing a network of more efficient lighthouses, so his trainees were developing the shipping which would rely on the lights. Robert Napier, son of a Dumbarton blacksmith, was one such, leaving the west coast to work for Stevenson at Edinburgh (see p. 9).

Civil Engineering

$\mathcal{M}$UCH of the shipbuilding innovation in Scotland and elsewhere depended on the breakthrough in design provided by James Watt with his modified steam engine, and of course the engine was used for locomotion and power for industry. Thanks to his diaries we are able to pinpoint what many would say is the beginning of the industrial revolution, to a precise Sunday in 1765, during a walk on Glasgow Green, where a stone now marks the spot.

Watt

James Watt (1736-1819) was born in Greenock, where he was to learn many basic engineering skills in his father's workshop. He trained in Glasgow as a mathematical instrument-maker and spent a short period in London before returning to follow this profession in a post at the University of Glasgow. When he married in 1764, Watt moved to a house in Delftfield Lane, the site of the Delftfield Pottery, of which he had been a partner for a few years. Later, when the pottery had closed, this site was used as an iron foundry by Boulton & Watt. Delftfield Lane was widened in the 1850s and renamed James Watt Street.

James Watt

It was in the cellar of his workshop, in the same year of 1764 that Watt repaired the model of the Newcomen engine which belonged to the university. This useful but inefficient engine was to inspire Watt's great contribution to industry. He hit upon the separate condenser for the conversion of used steam back to water. Together with other improvements to the efficiency of the engine, Watt was able to develop the usefulness of the invention as a source of energy to compete with water power. This took years, and it was only following his partnership with Matthew Boulton of Birmingham in 1774 that the engine became a commercial success. Patented in 1769, the engine was not put into production until 1774, at the Soho Engineering Works near Birmingham. Early references to Watt's invention describe it as the fire engine. In the 1780s the firm began to create the wealth that enabled Watt to retire, leaving the company in the hands of both his and Boulton's sons. Until that time Watt continued inventing and patenting refinements to the engine design, including the sun and planet motion, the parallel motion and the governor.

Watt's inventiveness was not restricted to these most famous creations. When he moved to Birmingham he was introduced to Josiah Wedgwood, and Watt later advised him on the pottery processes he had earlier overseen at Glasgow, sending him samples of 'Scotch clay' with notes on the firing and resultant finishes of the materials from the results of his own trials.

He was also employed on the surveys of the Caledonian and Forth & Clyde Canals and in the deepening of rivers, including the Forth and the Clyde. He was therefore responsible in various ways for the growth of Glasgow both as an city of industry and as a major port well inland from the previously navigable limits of the shallow River Clyde. In a career based on observing the behaviour of the element, Watt also discovered the composition of water.

He devised and constructed a machine for copying

manuscripts, and in his later years worked on a sculpturing machine, harking back to his days at the Glasgow pottery. He gave his name to the Watt unit of power, and also defined the formula for, and coined the term, horsepower – the rate at which work is done when 33,000 lbs are raised one foot in one minute.

James Watt's steam engine enabled the means of production to become much more flexible. Whereas water power demanded that the factory be located near the power source, steam engines could be placed anywhere. And steam engines could be run at any time of the day, without waiting for a heavy fall of rain, as was often the case with water-power.

Gas

William Murdock (1754-1839), born at Auchinleck, trained with his millwright father before moving south to work for Boulton & Watt. He made improvements to Watt's engine and invented various others including a steam-gun, before looking at the possibilities presented by the production of coal gas. He is most famous for his pioneering use of coal gas for lighting, which he used to light his house in 1792, although it was more than a decade before Boulton's engineering works were similarly lit.

David B Peebles (1826-99) was born at Dundee, where he served his time as an engineering apprentice. He moved to

Swindon's railway workshops before coming home, to a gas-meter manufacturer in Edinburgh. He saw the prospect of future growth in the gas industry and set up his own company, inventing various devices for the gas industry before branching out into electrical engineering. The company, Bruce Peebles & Co, achieved world renown for the producing transformers and rectifiers for electrical industries.

Electricity

James Clerk Maxwell (1831-79), mathematician and physicist, contributed to the field of study of electro-magnetism, leading to the later development of quantum physics.

George Forbes (1849-1936) was born in Edinburgh, the son of James David Forbes, Professor of Natural Philosophy there. His interests and innovations were diverse. While Professor of Experimental Philosophy at Anderson's College in Glasgow, Forbes travelled widely, including an expedition in 1874 to Hawaii. With James Young, Forbes improved methods of measuring the velocity of light. In 1880 he suggested the existence of a ninth planet in the solar system, anticipating the discovery of Pluto. His later years saw him working as an electrical engineer. His work on the design of dynamos included the invention of the carbon brush.

Sir **William Ramsay**, (1852-1916), the discoverer of the noble gases, was born at Glasgow. With Lord John Rayleigh, in 1894 he discovered argon, then helium and later other inert gases to which he gave the names neon, krypton and xenon. He was awarded the Nobel Prize for Chemistry in 1904.

Sir **James Swinburne** (1858-1958) was born in Inverness and began his working life at Manchester in a locomotive works, later joining R E B Compton's electrical engineering firm. It was here he invented the hedgehog transformer for wirelesses. He set up as a consultant and diversified into plastics but when he applied for a patent for a process making synthetic resin he was beaten to it, by a day, by Leo Baekeland, whose Bakelite became a household word. The two eventually came to an agreement to join production under the company name Bakelite UK Ltd. President of the Institution of Electrical Engineer and of the Plastics Institute, Swinburne had over 100 ideas patented.

Sir **Keith Elphinstone** (1864-1944) was born at Musselburgh, and his career in electrical engineering saw him devise and instal many innovations. He gained experience working in the earliest installations of telephones and electric light in London, and in 1893 became a partner in Elliott Brothers of London. The firm of electrical and mechanical engineers made the first micrometers, and Elphinstone devised the first continuous-roll strip chart recorder. At Brooklands motor-racing circuit he installed the original speed recording system to his own design, and designed a speedometer for private cars.

Charles T R Wilson (1869-1959), born at Glencorse near Edinburgh, was educated at Manchester and Cambridge. In the Second World War a practical application of his research into atmospheric electricity was the protection of barrage balloons from lightning. He devised the cloud chamber method of tracking alpha-particles and electrons, which allowed the movement of atoms to be recorded on film. He was joint winner with Arthur Compton of the Nobel Prize for Physics in 1927.

Power

$\mathcal{T}$HE pioneers of shipbuilding on the Clyde passed on a lasting legacy of engineering excellence to those who followed.

Steamships

Patrick Miller (1731-1815) patented the paddle wheel for early steam boats. A merchant banker, in 1788 Miller combined forces with **William Symington** (1763-1831) of Leadhills, launching a steam boat with paddle-wheels between its twin hulls, on Dalswinton Loch in Miller's Dumfriesshire estate. James Watt was not enthusiastic about this design, and Symington's *Charlotte Dundas*, the first practical steam boat, was even less well received. Vested interests combined to claim that the wash from the steam boat would damage the banks of the Forth and Clyde Canal. Despite this first, with a patented engine providing direct drive to the paddles, and his patent in 1787 for a road locomotive engine, Symington died in poverty.

In America a counter-claim for the world's first steam boat demonstration is made for **James Rumsey** (b1754), born to Scots parents in Virginia. His boat made its trip in 1786 on the Potomac River in the presence of George Washington.

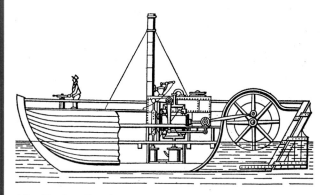

The Charlotte Dundas

SOME SCOTTISH SHIPBUILDING FIRSTS

- the first paddle steamer
- the first iron steamship
- the first all steel ship
- the first paddle steamer to cross the Atlantic
- the first ship to cross the Atlantic in less than a week
- the first all-welded ship
- the first merchant ship to run on oil
- the first set of triple-expansion engines for a twin-screw steamer
- the first ship to be fitted with two engines
- the first steam whaler
- the first steel ship to cross the Atlantic

Robert Fulton (1765-1815) is said to have seen the first sailing of the *Charlotte Dundas* in 1801. Two years later Fulton was cheered by a Parisian crowd at the sight of his steamboat on the Seine. In 1807 he launched America's first practical steamboat, the *Clermont*, on the Hudson, with financial help from Robert Livingston. That same year the pair established the world's first commercial steamboat service, between New York and Albany.

Henry Bell (1767-1830) trained as a wheel-wright and ship-builder. He worked in London before returning north where his work as an engineer at Helensburgh led to the launch on the Clyde in 1812 of *The Comet*. Europe's first passenger steamboat, it sailed regularly between Greenock and Glasgow .

Robert Wilson (1803-82) devised the first practical screw pro-peller, the alternative to the paddle. Sir William Fairbairn made the first riveting machine for metal. He built the first iron steamship, the *Lord Dundas*, at his Millwall yard.

The Comet

Marine Engineering Innovations

John Scott Russell (1808-82) turned to naval architecture after his steam passenger coach, en route from Glasgow to Paisley, crashed with four fatalities. His experiments on hull construction, carried out on canals, lead to the 'wave line' form which reduced drag. He discovered the solitary wave, which he named the soliton, caused by a canal boat stopping suddenly. He followed the single wave on horseback for some distance, and the mathematical breakthrough engendered by this observation is today being utilised in the world of fibre optics.

James Howden (1832-1913) set up his own business to make marine engines and boilers in 1854, later going on to establish James Howden & Co in 1862. He invented the forced draught system for marine engines, improving efficiency.

James Graham, 6th Duke of Montrose (1878-1954) was born in London. As a director of William Beardmore & Co in Glasgow he designed one of the first sea-going vessels with a heavy oil engine, the *Mairi*. His main innovative credit is his 1912 plan for an aircraft-carrier.

*T*HE surge in creativity in the late 18th and early 19th centuries brought huge changes to society. It clearly was a revolution, with people moving off the land and into the cities where factories were springing up.

Economics

Adam Smith (1723-90), the renowned Kirkcaldy-born economist and philosopher made a huge contribution to the development of the industrial age with his *Inquiry Into the Nature and Causes of the Wealth of Nations*. He examined the roots and consequences of the market economy, and suggested the division of labour. Observing the production of pins, he concluded that the whole process would be speeded up by improved efficiency if divided into individual tasks. Each worker would carry out one task rather than the whole production, in a precursor of the production line famously used by Henry Ford over a century later.

Iron Industry

David Mushet (1772-1847) opposed popular opinion to show that the black-band ironstone he found at William Dixon's Calder Ironworks was suitable for smelting, allowing it to become widely used in later years. He also demonstrated that the quality of wrought iron could be improved by the use of non-phosphoric oxides of iron and that iron and steel could be improved by adding manganese. He patented a process using wrought iron to make cast steel.

John C Loudon (1783-1843) devoted his life to the study of horticulture. His practical invention of a wrought-iron sash bar capable of holding its strength and shape while bent in any direction was the foundation of the designs of glass-houses and winter gardens so popular in the Victorian age, culminating in the Crystal Palace designed by Paxton for the Great Exhibition.

James Beaumont Neilson (1792-1865) invented the hot blast oven, a revolutionary invention for the iron industry which he

Heavy Industry

The Winter Gardens, Glasgow made possible by John Louden's advances with wrought iron

experimented with at 'Dixon's Blazes' works south of Glasgow's Gorbals area. The process was more efficient than the method used until then, reducing the amount of coal required in the production of iron. The growing rail and shipbuilding industries were to benefit in turn, as Glasgow became known as the 'Workshop of Empire', building goods and machines for export throughout Britain's foreign dependencies and beyond. Neilson also established the Glasgow Gas Works promoting the use of gas for lighting. He invented the swallow-tail burner, described at the time as 'a beautiful, smokeless and economical light', and was soon supplying gas lighting for streets and indoor use.

Manufacturing

Sir **Isaac Holden** (1807-97), a Liberal MP from Renfrewshire, made his fortune from the manufacture of woolcombing machinery, only after a hefty investment in its development – around £50,000, a huge total in the mid 19th century.

James Nasmyth (1808-90) was the youngest son of artist Alexander Nasmyth. Alexander took part in Patrick Miller's early paddle steamer projects (see p. 28). James entered a career

in the sciences, starting in the foundry business, where he devised a steam hammer to forge a wrought-iron paddle shaft. This was his most famous invention, but he also devised a planing machine, pile-driver and steam lathe. Taking his interest in astronomy further than most amateurs, he designed and made his own telescope using the 'Nasmyth focus', a design revisited in the next century.

Robert Francis Fairlie (1831-85) trained at the railway centres of Crewe and Swindon. He established his own consulting engineering business in London, patenting a double-bogey engine for use on narrow-gauge railways for carrying freight.

Fairlie locomotives became widely used, and can be seen today in Wales at the Ffestiniog Railway.

Sir **Dugald Clerk** (1854-1932) invented the two-stroke Clerk Cycle Gas Engine in 1877. His specialist knowledge of internal combustion engines gave him a leading research post during the First World War.

Dugald Clerk

SCOTTISH AGRICULTURAL INVENTIONS & INVENTORS

- Sir **Hugh Dalrymple** (1700-1753), Lord Drummore: invented hollow-pipe drainage, making previously unworkable water-logged land usable.

- **James Anderson** (1739-1808) invented the so-called 'Scotch Plough', a two-horse plough with no wheels.

- Clergyman **Patrick Bell** (1799-1869) invented a mechanical reaper – a flop at home, but widely used in America.

- **James Meikle** (c1690-c1718) of Dunbar, East Lothian invented a fanning device for winnowing grain (separating grain from chaff); he also invented a barley mill.

- **Andrew Meikle** (1719-1811), son of James, devised a threshing machine adaptable to wind, water, horse or steam power; he also redesigned windmills.

- **James Smith** (1789-1850) invented a soil-drainage system, named 'Deanstonisation' after the Perthshire cotton mill where he worked; his other inventions included reaping machines and a salmon-ladder.

- **Graham Tuley** a Scot by adoption, invented the award-winning Tuley Tree Shelter a synthetic sleeve to protect vulnerable saplings and provide them with their own growth-promoting micro-climate.

- A Scots first: Sir **William Hooker** founded Kew Botanical Gardens.

*T*HE two major means of communication today, whether at home or in business, are TV and the telephone, although these are still run a close third by the postal service. The main attribute of these methods of communicating is the connection of people over long distances. This is the main benefit and the crucial element which has made them so important, and Scots were leaders in each. But before these inventions came to be, there were many pioneering Scots at work in other methods of making contact or transmitting ideas.

Stereotyping

William Ged (1690-1749) was an Edinburgh goldsmith who patented a stereotyping process in 1725. Commissioned by Cambridge University to stereotype prayer books and bibles, he met with opposition from workers and, in common with so many innovators, died in poverty. In this context stereo, meaning solid or three-dimensional, applies to metal poured into a mould to create a block which can be used and re-used for printing. His 'lost wax' method of metal casting proved especially useful for the reproduction of fine and delicate designs, particularly effective in jewellery making.

Postal Innovations

Son of the manse **John Anderson** (1726-96) was born in Rosneath in Dunbartonshire. He was Professor of Oriental Languages and of Natural Philosophy at the University of Glasgow, and allowed working men to attend a class in mechanics, in their working clothes. After his death his estate was used to establish Anderson's College, which later became Strathclyde University. His contribution to the field of communications was the invention of the balloon post.

James Chalmers (1782-1853) from Arbroath, worked as a bookseller in Dundee when, in promotion of a more efficient postal system, he invented the adhesive postage stamp in 1834.

He was also responsible for inventing the postmark, cancelling his first invention from being used an illicit second time, and identifying the place of posting. Scotland was to see the first examples of other elements of the mail service in the UK, with the first post office, at Sanquhar in 1783, and the first motor mail-van service in, Argyllshire in 1896.

Signalling

Philip H Colomb (1831-99) served in the navy in the Burmese War and in China. He recognised the importance of naval supremacy, and designed a system of signalling at night named 'Colomb's Flashing Signals.'

Printing

George Wilson was the world's first Professor of Technology, at Edinburgh University, where he studied uses of colour and the problems of colour blindness. He was the first to suggest that recruiting for jobs such as railway signalmen should be include testing for the condition. Consulted in the 1850s by Bank of Scotland directors on their concerns over forgery, he recommended the use of different colours for different denominations of bank note.

Alexander Crum Brown (1838-1922), Professor of Chemistry at Edinburgh University, took the idea a step further, proposing that to overcome the photographer's ability to reproduce images, the Bank should use inks composed of the same substances in different proportions, making them more difficult to fake.

Time-keeping

The standardisation of time measurement and synchronisation has been essential to the communications industry. Sir **Sandford Fleming** (1827-1915), born in Kirkcaldy, was a railway engineer who made his name in Canada. Chief Engineer of the Canadian Pacific Railway, he surveyed many of the country's rail routes. He devised an internationally recognised system of standard time in 1884.

Telegraph

In 1753, Charles Morrison wrote to the *Scots Magazine*, suggesting a method of constructing the first practical telegraph. Many Scots took the invention further, but it was American Samuel Morse who constructed the first practical telegraph some 90 years later. The ability to communicate over long distances became the aim of many inventors, with diverse results.

James B Lindsay (1799-1862) was born in Carmyllie in Forfarshire. He left the weaving trade, becoming a lecturer in the Watt Institution in Dundee, and pioneered the uses of electricity. His many inventions included an electric telegraph and a continuous electric light, and he demonstrated wireless telegraphy through water in furtherance of his suggestion that a transatlantic telegraph was feasible.

Alexander Bain (1810-77) of Watten in Caithness trained at Wick as a clockmaker, leaving for London in 1837. An innovator in the manufacture of clocks, he patented a design using electro-magnetic pulses, leading to his proposal that many clocks could be synchronised to a central time. He patented a device for the electrical transmission and chemical recording of messages and images. The device foresaw the development of the fax. Returning to Edinburgh, he developed the first practical telegraph recording device. Bain died in poverty but it was a Scot who took his invention further.

Teleprinter

Frederick G Creed (1871-1957) was born in Nova Scotia of Scottish parents, and in his teens worked as a telegraph operator. He travelled extensively before settling in Glasgow. There, his experiments with recording messages on perforated tape led to his invention of the teleprinter. He initially manufactured the machines at Croydon, but later sold the business to ITT.

Telephone

Alexander Graham Bell (1847-1922) was born in Edinburgh into a family whose professional lives were devoted to communication. His father and grandfather both taught elocution, and Bell was to give much of his time to the improvement of communication techniques for the deaf and mute. It may seem ironic in the light of his most famous invention, but the research which led to the telephone was initially aimed at finding a means of remote communication for the deaf.

The harsh Scottish climate drove Bell to Canada and then America in search of a home more comforting to his health. In Boston he was made Professor of Vocal Physiology, devoting his time to the teaching of deaf mutes, using 'visible speech', a system devised by his father, Alexander Melville Bell. This involved the illustration of the positions of the vocal chords for making each sound. Experimenting with acoustics led Bell to attempt to send messages which could be 'heard' by deaf people. Ultimately this brought the first intelligible telephone message, on the 5th of June, 1875. In the following year Bell patented the telephone, and formed the Bell Telephone Company in 1877. Bell's most famous pupil was Helen Keller, the deaf and blind woman renowned for her work with the handicapped. He married another of his pupils, who was also deaf. It seems Bell disliked the telephone itself, finding it noisy and intrusive.

Wireless

Sir **James Swinburne** (1858-1958) was born at Inverness. He moved to Manchester, where he trained as an electrical engineer. Towards the end of the 19th century he invented the hedgehog transformer for wireless sets, and it is in the manufacture of wireless sets that his pioneering work in plastics is most seen. Together with rival Leo Baekeland he formed Bakelite Ltd in 1926 (see p. 27).

Television

John Logie Baird (1888-1946) of Helensburgh is recognised as the father of this medium. In common with many of his fellow inventors, Baird was a son of the manse. Indeed he claimed that his classical education turned him away from the career in the ministry his father hoped he would follow; his rebellious streak ushered him towards more practical work even at an early age. Interested in the new technology of the telephone, the twelve-year-old Baird connected his house to those of a group of friends with their own telephone system. This unofficial relay was closed down when a passing carriage-driver was thrown from his seat by a low-slung cable. As a Glasgow University student Baird's subject was electrical engineering. Poor health

John Logie Baird

interrupted his career, and led to him invent the Baird Undersock, a garment worn inside ordinary socks to keep feet warm and dry. This was financially rewarding and its marketing included the first sandwich-board women seen in Glasgow and a host of friends visiting shops to ask for the undersocks.

Settling in England, Baird began experimenting in the technology of television. Knowing that he had competition, he divided the work into smaller projects and set others to work on the component parts, each unaware of the larger scheme to which they were contributing.

His first public demonstration was just ahead of the competition, in January 1926. His original 'contraption', made with an ad-hoc arrangement of darning needles, hat boxes, a biscuit tin, a tea chest and electric motors stuck with sealing wax and glue, may in fact have been a piece of subterfuge designed to put rivals falsely at ease. In fact the Bell Company made a public broadcast little more than a year after Baird's first demonstration. He followed up with a broadcast from Glasgow to London, and then from one side of the Atlantic to the other.

The British Broadcasting Company (BBC), then with a monopoly on public broadcasting, and until then only in radio, initially chose Baird's technology for their own use. But they later changed, favouring the Marconi-EMI system.

Baird went on to look for further innovations in the field of television. He showed the possibility of broadcasting in colour, 3D broadcasts, and recording television programmes on magnetic disks. He took out a patent on fibre-optics and patented a design for night-vision optics using infra-red. Although not publicised at the time for obvious reasons, it seems he also did pioneering wartime work on radar.

Radar

Sir **Robert Watson-Watt** (1892-1973) was born in Brechin. Descended from James Watt, he was a graduate of the University College, Dundee, where he then worked as a researcher. In the widening research required with the growing aviation industry, Watson-Watt found a place in the meteorology department of the Royal Aircraft Factory at Farnborough. His work looking at the detection of inclement weather was directly for aviators in the First World War. The success of this research led to the appointment in the inter-war years in charge of the radio department at the

Robert Watson-Watt

National Physical Laboratory. The purpose of the department was to develop radar. As with Baird and television, Watson-Watt was aware of others working in the same area of research, but with the government's belief that a radar system would be have significant defence benefits the British kept ahead of the Germans. This was proved crucially during the Battle of Britain in 1940, when control of the skies repelled a potential invasion.

After the war the peacetime application of radar was crucial to the growth of civil air travel, and Watson-Watt was recognised across the world as the pioneer.

Alan Archibald Swinton (1863-1930), like Baird, was still a teenager when he linked his house with that of a friend by way of Bell's new invention, the telephone. As an electrical engineer he used lead to insulate a ship's electric cables, an innovation in the 1880s. His work as a consulting engineer led him to look into the medical uses of radiography. In 1908 he described the concept and basis of electronic television, using magnetically reflected cathode ray tubes at both the camera and receiver.

SCOTTISH PUBLISHING FIRSTS

- the first book translated from English to a foreign language
- the first edition of the *Encyclopaedia Britannica* (1768-81)
- the first English textbook on surgery (Peter Lowe, 1597)
- the first modern pharmacopaedia, the *Materia Medica Catalogue* (William Cullen, 1776)
- the first book textbook of Newtonian science (David Gregory)
- the first Bible in raised type for the blind
- the first colour newspaper advertisement
- the first postcards and picture postcards in the UK

SCOTTISH CHEMICAL INNOVATIONS

- **William Cullen** (1710-90) of Hamilton was largely responsible for establishing chemistry as a distinct discipline, with practical applications in agriculture and industry.

- **Archibald Cochrane** (1749-1831), 9th Earl of Dundonald, invented several patent methods of manufacturing chemicals, experimenting on his Culross estate in the manufacture of coal tar, British gum, white lead and alkali.

- Sir **Charles Tennant** (1768-1838) of Ochiltree, Ayrshire, was an industrial chemical baron after he patented a manufacturing process for a dry bleaching powder in 1799. The benefits of an easily transportable bleaching agent made his St Rollox the largest chemical works in the world.

- **Thomas Graham** (1805-69) of Glasgow was called 'the father of colloid chemistry'. He devised dialysis, a method for separating particles into two classes – colloids (such as gum arabic, with low diffusibility) and crystalloids (such as common salt, with high diffusibility).

- **William Gregory** (1805-58), from a long line of Edinburgh scientists, successfully isolated natural substances, preparing morphine and codeine from opium and making isoprene, a basis for synthetic ubber, from crude rubber.

Chemistry

SCOTTISH CHEMICAL INNOVATIONS

- **James Young** (1811-83) from Glasgow established the Scottish mineral oil industry, distilling bituminous substances extracted from coal and shale. He produced lubricating oils and oils for lighting, earning the nickname 'Paraffin' Young.

- Leith-born **Thomas Anderson** (1819-74) discovered picoline and the pyridine derivatives of oil and coal-tar. Another of his discoveries, anthracene, used in dyeing, is also a derivative of coal-tar.

- **Archibald S Couper** (1831-92) of Kirkintilloch, produced fundamental theories in the formulation of organic compounds.

- Shettleston-born **William Cullen** (1867-1948) worked for the Nobel Explosives Company in South Africa, developing smokeless explosive powders.

- **George Burnett** (1921-80) led the science of polymer chemistry, establishing quantifiable processes of polymerisation in a previously experimental field.

- **Thomas Stevens** (b1900) of Renfrew has had three chemical reactions named after him. The most famous, the Stevens rearrangement, is a reaction where the molecular structure of an ion carrying both positive and negative electric charges (known as an organic zwitterion) is rearranged to form a different compound with the same chemical formula.

Chemistry

$\mathcal{M}$EDICINE has traditionally been one of Scotland's strengths, and the Scots' contribution to medicine has been enormous, with the innovation still continuing today.

Chemistry

Hamilton-born **William Cullen** (1710-90) while a pioneer in creating chemistry as a field of study in its own right (see p. 43), was also the main publicising force in showing the place of the nervous system in the health and illness of the body.

Joseph Black (1728-99) was Professor of Anatomy and Chemistry at Glasgow University, then Professor of Medicine and Chemistry at Edinburgh. Thought of as the Father of Quantitative Chemistry, his innovations include the concept of 'latent heat'. This is the heat required to change a solid to liquid or liquid to gas, without a change in temperature, and was to help Watt in the development of the steam engine. He also discovered 'fixed air' or carbon dioxide and established the lack of carbon dioxide in lime and alkalis, which is contained in limestone and carbonates of alkalis. This crucial difference explains the causticity of the former, and established the notion that a gas can exist in a solid.

Botany

Robert Brown (1773-1858), although primarily a botanist, contributed to the world of physics in general as a result of his investigations into the reproduction of plants. The Montrose-born son of an Episcopalian minister, he became a leading botanist after military service. He brought home more than 3000 species of plant life from Australia, and was keeper of Sir Joseph Banks' collections. He was the first to identify and name the nucleus in living cells.

Hypnosis

James Braid (1795-1860) was a surgeon who pioneered the

Medicine

field of Hypnosis, coining the term Neurohypnosis, which was later shortened to hypnosis.

Homeopathy

Robert E Dudgeon (1820-1904) was born at Leith. A medical graduate of the University of Edinburgh, he studied overseas, then settled in Liverpool. An early proponent of homeopathic medicine, in 1878 he invented a device for pulse-rate measurement which was named after him.

Physiology

Alexander Crum Brown (1838-1922), born in Edinburgh, is remembered in the Crum Chair of Chemistry at Edinburgh University, where he was Professor of Chemistry from 1863 until his death. He devised the diagrammatic system of representing chemicals compounds, with connecting lines showing bonds, a system still in use. His important work as a physiologist related to the connection between vertigo and the sense of balance in the workings of the inner ear.

Cancer Medicine

Sir **Alexander Haddow** (1906-76) of Broxburn advanced cancer treatment by identifying the similarity between carcinogens and substances which can kill cancerous and healthy cells.

Prenatal Ultrasound

Ian Donald (1910-87) was the son of a Paisley doctor. An obstetrician by profession, he used echo-sounding techniques with radar technology to make the first ultrasound scanner. With engineer Tom Brown he developed the first machine from its origin in 1958 to the scanners used as a matter of course by the 1970s in Glasgow. Innovative research continues at the Scottish universities.

Pharmacology

The difficulties presented by large dosages of drugs have been the downside of the ability to operate on many conditions. As anaesthetics had to work as muscle relaxants a high dosage was required, resulting in potentially dangerous effects on the patient's system and a long period of recovery. Modern relaxants have been developed to reduce the need for large doses by concentrating their effects on more localised areas. Scottish Universities have played a large part in the development of such drugs, working in combination with commercial pharmaceutical companies. Two muscle relaxants were created by teams at Strathclyde University: atracurium with Wellcome, and vecuronium for Organon. The former was developed from the arrow poison used by indigenous people of South America, and since its license was granted in the late 1970s has earned over £10m in royalties for the University.

Genetics

Struther Arnott (b1934) of Larkhall was educated at Glasgow University and spent time at King's College London and Purdue University, Indiana before returning to St Andrews University in the 1980s. His research concentrated on chemical structures and methods of illustrating the results. He discovered new conformations of DNA, valuable in furthering the pioneering genetic research which is carried out extensively in Scotland.

Magnetic Resonance Imaging

John Mallard of Aberdeen University looked at a new way of scanning the body for potential problems. Professor Mallard's MRI scanner, developed in the 1980s, was based on the magnetic resonance imaging of the body, which, by illustrating the presence of water, shows the difference between healthy and diseased tissue.

Medicine

Orthopaedics

A team of 'cell engineers' at Glasgow University has developed a new healing process for injured tendons. The 'smart splint' is a means of repairing torn tendon ends without deformity of the joint: by introducing biodegradable materials containing tiny grooves to the tissue, the tendon will grow together in a continuous form.

Besides these innovations, there are various areas of medicine which owe much of their development to Scots. These might be categorised as preventive, curative and surgical.

Preventive Medicine

John Arbuthnot (1667-1735) of Kincardineshire was a leading society character in early 18th London. As physician-in-ordinary to Queen Anne he was a doctor of renown, and numbered many literary celebrities among his wide circle of friends. In his satirical writings *The History of John Bull*, he created the typical Englishman in the image of his eponymous hero. Professionally, he was a man of vision, making the point in an essay in 1731 that diet is a major factor in the treatment of disease.

James Lind (1716-94) was born at Edinburgh. He earned the title of the 'father of naval hygiene' with his improvement of health among seamen. Following naval service he became physician at the naval hospital at Gosport. There he made the prevention of scurvy a priority, and in 1795 he succeeded in persuading the Admiralty to introduce lemon juice to the men's diet, many years after his *A Treatise on Scurvy* was published (1753).

William P Alison (1790-1859) was born at Edinburgh, and educated there, becoming physician to the New Town Dispensary. There he became concerned with the health and welfare of the poorer members of society, particularly with regard to the epidemic diseases suffered in the cities of Scotland at the time. He believed there was a connection between living conditions and the worsening health of the populace. He produced a report, *Observations on the Management of the Poor in Scotland, and its Effect on the Health of the Great Towns*, in 1840. His criticism in the report of the system of poor relief provoked a reaction in Government, putting many of his ideas to work in an Act of Parliament in 1845. As a proponent of Government intervention in health and welfare of the country's citizens generally, his ideas are seen as a forerunner of the National Health Service.

Sir **Henry Littlejohn** (1826-1914), born at Edinburgh and educated in Scotland and at the Sorbonne, gained an international reputation for his work relating to sanitation and public health. Trained as a surgeon, he lectured for many years but it was his practical work as the first Medical Officer of Health for Edinburgh which gave him the means of reforming the city's health. Through his pressure and his reportage of the sanitary conditions of the capital, Edinburgh became the first city to obtain an Act of Parliament making the notification of every case of infectious disease compulsory.

Sir **Dugald Baird** (1899-1986), like William Alison before him, had his ideas on social welfare shaped by working among the city slums. As a student he saw at first hand the poor conditions in which the poor women of Glasgow gave birth at home. As Regius Professor of Midwifery at Aberdeen University he had the opportunity to shape government policy with research into the higher maternal and infant mortality rates of the poor. By combining the work of dieticians, statisticians and sociologists he demonstrated improvement in the health of the Grampian population and taught his methods to promote what he called the Fifth Freedom for women – the freedom from excessive fertility. He was, therefore, also a supporter of reform of abortion law in the 1960s.

Medicine

John Boyd Orr

Research into the mineral content of soils and pasture and the diet of livestock convinced **John Boyd Orr** (1880-1971) that nutrition of the poor could be improved. He embarked on literally ground-breaking research as the Director of the Aberdeen University Institute of Nutrition. Accepting financial assistance from his old school friend John Quilter Rowett, he created the now world-famous Rowett Research Institute. He demonstrated the connection between the mineral constituents of pasture land and the nutrition of grazing animals, showing that illness was sometimes due to mineral deficiencies. Persuading farmers to adopt provide good nutrition for their livestock, Boyd Orr saw the potential rewards for the poorer members of society. He demonstrated a marked improvement in growth of poor children given milk, but his study gave the unpalatable message to the Government that less than half the British population earned enough for an adequate diet. After trying to challenge Boyd Orr's statistics, the Government introduced the free-milk scheme for schoolchildren. Taking a wider view, he showed that while half the world starves, farmers face lower prices if they over-produce. He concluded that intervention was required to encourage farmers to produce more without loss of income. For his work to achieve a world without hunger he received the 1949 Nobel Peace Prize.

Medicine

Curative Medicine

Many treatments of medical conditions do not involve invasive techniques. These non-surgical treatments involve the identification of antibodies, the use of antibiotics, and the prescription of pills, tablets and creams. Many Scots have brought us from the days when it was believed that health was governed by the four humours.

George Cleghorn (1716-94) was an army surgeon when he discovered that malaria can be cured by use of quinine bark.

Sir **Patrick Manson** (1844-1922), born at Old Meldrum in Aberdeenshire, was a pioneer of tropical medicine. He showed that the mosquito is the carrier of malaria, and led research on sleeping sickness and beri-beri.

Concluding that research was Sir **David Bruce** (1855-1931), an Australian of Scottish descent, who discovered that the tsetse fly was the carrier of sleeping sickness. He isolated the bacteria of Malta fever and it was named brucellosis after him.

Sir **Ronald Ross** (1857-1932) was awarded the Nobel Prize for Physiology or Medicine in 1902 after his work confirmed Manson's by showing how malaria will spread and describing how it can be combated in *Prevention of Malaria.*

A Canadian team including the Scot **John J R Macleod** (1876-1935) together with Sir Frederick Grant Banting and Charles Best completed pioneering research into the control of diabetes. Banting and Macleod shared the Nobel Prize for Physiology or Medicine in 1923 for the discovery of insulin.

Perhaps the greatest contribution to the non-invasive treatment of ill-health has been that of Sir **Alexander Fleming** (1881-1955). So complete is the general acceptance of his discovery that any mouldy food is likely to provoke a joke about penicillin. Born at Lochfield in Ayrshire, the son of a farmer, he was an outstanding student at St Mary's in Paddington,

Medicine

Alexander Fleming

London, graduating in 1906. The First World War saw Fleming join the Royal Army Medical Corps, and his experience of treating wounds was to develop his earlier interest in bacteriology. As a researcher he was the first to administer an anti-typhoid vaccination to humans, and during the First World War he discovered an antiseptic element present in tears. In 1928 he famously returned to work to discover mould on a lab culture. He crucially noticed that the bacteria around the mould had been destroyed, and identified the common mould as penicillum. His development of penicillin as a drug was limited, and it was not until the early 1940s that others were able to produce the drug commercially. At this time all penicillin produced was given over for sole use by the armed forces, helping counteract the effects Fleming had seen at first hand in the First World War. The fame of the discoverer is matched by the effectiveness of this early antibiotic which, 50 years after its discovery, is still the treatment of choice for a wide range of infections, from minor ear complaints to potential killers like meningitis.

Sir **William B Leishman** (1865-1926) was a Glasgow-born bacteriologist discovered an effective vaccine for typhoid. He was also identified the protozoan parasite which carries the tropical fever, kala-azar.

Sir **John W Crofton** (b1912) researched chemotherapy of tuberculosis, demonstrating that bed rest was not an essential part of treating the disease. The result was the closure of TB hospitals here and abroad. The wider application of other aspects of his findings have resulted in the successful treatment of TB in many countries.

Alick Isaacs (1921-67) first isolated interferon during research into the behaviour of flu viruses.

Sir **James W Black** (b1924) of Beath was educated at Cowdenbeath and St Andrews University. He has lectured in Malaya, Scotland and England, and worked in research with ICI Pharmaceuticals, Smith, Kline and French and Wellcome Research Laboratories. Knighted in 1981, he won the Nobel Prize for his work for ICI, where he was among the team that developed the first beta-blocking drug to have clinical purpose. At Smith, Kline and French he developed the drug cimetidine.

Surgical Medicine

Early surgeons had to work with great speed as well as accuracy, without the benefit of anaesthetics and antiseptics. This also made many conditions inoperable, leading people to misunderstand the nature of the diseases.

The **Gregory** family have been held up as an example of hereditary genius. This Aberdeenshire family produced 16 professors over five generations, and gave its name to one of the most prescribed drugs of its day, Gregory Powder, more popularly known as Gregory's Mixture. Used against stomach upsets, it was a mixture of magnesium carbonate with powdered rhubarb and ginger, invented by **James Gregory** (1753-1821). Another family member, **William Gregory** (1803-58), experimented widely in the preparation of compounds which could relieve pain. Some ranked him with Thomas Graham as the only British chemists of international distinction. Gregory perfected the manufacture of chloroform and morphia, and was the first to prepare morphine hydrochloride in pure crystalline form, making it a practical form of pain relief.

Robert Liston (1794-1847) was born at Ecclesmachan Manse in Linlithgowshire. A skillful surgeon of wide renown, he was the first surgeon to use a general anaesthetic in a public operation, in 1846. The Liston splint, which he invented for treatment of dislocated thigh, is named after him.

Sir **James Young Simpson** (1811-70), born in Bathgate, the son of a baker, but with his family's support at 14 years old he went to Edinburgh University. Specialising in obstetrics, he became Professor of Midwifery in 1840. He was the first to use ether as an anaesthetic in childbirth. He used himself and his assistants as guinea pigs in the search for a better anaesthetic and, finding chloroform to meet his requirements, was the first to operate using chloroform. He was opposed by the medical and religious establishment, but in persuading Queen Victoria of the benefits of chloroform for the birth of Prince Leopold in 1853 he achieved wider popularity for his methods. The aforementioned William Gregory helped make chloroform widely available, producing the drug on a large scale. He was the founder of modern gynaecology.

Joseph Lister (1827-1912) pioneered the use of antiseptics, so reducing the number of deaths from infection after operations.

Sir **William MacEwan** (1848-1924) was born, educated and worked in Glasgow. He made pioneering operations on the brain for the treatment of tumours, abscesses and trauma. A student while Joseph Lister was Regius Professor of Surgery, he later occupied the same position. Although renowned as a brain surgeon because of his ground-breaking work in this new field, he also performed many new techniques on other parts of the body. He devised ways of grafting bone to replace missing parts of bones in limbs, work of particular value during the First World War.

Norman M Dott (1897-1973) found his interest in medicine following a motoring accident which hospitalised him. He showed early inventiveness when, as a student, he was responsible for the endotracheal tube. Specialising in the field of neurosurgery he pioneered a successful treatment of aneurysm in the brain, using injected 'dye' to show up the problem under X-ray. Combining his neurological work with practice in paediatrics he also led the treatment by operation of congenital dislocated hips. He is considered, along with Sir Geoffrey Jefferson at Manchester and Sir Hugh Cairns at Oxford, to have established neurological surgery in Britain, after much founding experimentation and theoretical work on brain surgery had been carried out by the Aberdonian Sir David Ferrier.

David Ferrier (1843-1928) graduated in classics and philosophy at Aberdeen University before taking his MD at Edinburgh. Taking up the especially created Chair of Neuropathology at King's College, London, he identified by experimentation on monkeys and other vertebrates that brain functions are located in particular areas. This being the case, he deduced, brain tumours and injuries could be treated by operation.

Medicine

*T*HERE are many items in the house which have been invented and developed by Scots.

Paraffin

James Young (1811-83) was born in Glasgow. Although he initially followed his father's trade as a joiner, he attended evening classes in chemistry, eventually being appointed assistant to his professor. Young patented a means of making paraffin from coal, and at Bathgate manufactured naphtha and other oils before paraffin became sufficiently popular to produce in commercial quantities. The success of his discovery led to court cases to defend the patent, and Young was able in time to sell the manufactory for several hundred thousand pounds – a fortune in the 1860s – for paraffin lamps were then common household items.

Macintosh

Charles Macintosh (1766-1843) of Glasgow was the son of a chemical manufacturer and dyeworks owner, George Macintosh. He followed his father into the chemical industry, and was a partner in the Hurlet and Campsie Alum Works, producing chemicals for the dyeworks industry. It was at Hurlet that he experimented with solutions of rubber for waterproofing of fabrics. He obtained a patent for such a fabric in 1823, and the new textile was soon tailored to the rainwear that is still referred to as a 'Macintosh'. (Such was the fame of Charles Macintosh that a young Glasgow architect determined to use his middle name to avoid confusion; Charles Rennie Mackintosh also changed the spelling of his surname, adding a 'k' to the original.) It is said that the method of waterproofing patented by Macintosh had already been made known by another Scot, surgeon **James Syme** (1799-1870), in 1818. However Syme was more concerned with his career in clinical surgery and pathology, and it was the industrial chemist who developed the method commercially.

Matches

Sir **Isaac Holden** (1807-97), was born at Hurlet. He invented the Lucifer match, although it was his woolcombing machinery which brought his fortune (see p. 32).

Kaleidoscope

Sir **David Brewster** (1781-1868) of Jedburgh developed a wide interest in science while training for the ministry, ultimately opting at university to swap the philosophical for the practical. First as a science journalist and then in experimentation, his main interests were in optics. He invented new types of micrometer, for measuring minute distances, the dioptic lens used in lighthouses mentioned on page 21, and the stereoscope toy which used two images to give the impression of a solid or 3D picture. His best-known invention was the kaleidoscope, which caught on rapidly as a toy for all ages, although Brewster

envisaged it as a design aid for carpet manufacturers. Although he patented the device, his patent was insufficiently precise in definition. As a result it was widely copied, and the fortune Brewster imagined might have come his way, never did. However, his contribution to science as an inventor, journalist and founder of The British Association for the Advancement of Science, led to a knighthood in 1831.

Fountain Pen

Robert Thomson's (1822-73) earlier work on the development of a range of rubber products (see p. 15) had clearly stayed with him, and he used the material's adaptable properties when he designed and patented the principle of the fountain pen.

Piano with Foot Pedals

John Broadwood (1732-1812), born in the Borders village of Oldhamstocks, went to London in his late twenties to train as a harpsichord maker. Such was his skill that he eventually took over the company, helped along the way by marrying the proprietor's daughter. But his improvements to piano design meant he gave up making harpsichords, so popular was his new Grand Piano. The introduction of foot pedals, replacing the old hand and knee levers, was followed by improvements to the tone and range of the piano. His patent for 'piano and forte pedals' and the separate bass bridge were adopted by other makers. In 1811 he handed over the company to his sons and it continues today.

Lawnmowers

Although not the first to produce such a machine, **Alexander Shanks** of Arbroath patented his lawnmower design in 1842, a year after he first made a mower. His machine was pulled by a pony and was more effective than Edwin Budding's original, in that it left a smoother lawn. Horse-drawn lawnmowers were used until the mid 20th century, and the horses would wear overshoes to protect the grass. Shanks continued refining and developing his designs, but it was another Scot, **David Cockburn**, who designed the first rotary mower, the forerunner of today's domestic machines.

Vacuum Flask

Sir **James Dewar** (1847-1932) was born in Kincardine on Forth, the son of an innkeeper. As an experimental scientist he worked for many years on high-temperature physics before turning his attention to the liquification of oxygen and hydrogen, requiring low temperatures. The principle of vacuum insulation was not new to Dewar when in 1892 he required a vessel which would hold liquid gases and keep their temperatures stable. For this he developed the vacuum flask, and by refining the design to allow manufacture of flasks in robust materials, the invention went into commercial production in 1898. Although aware of the wider possibilities of a container which kept liquids hot or cold as required, Dewar did not patent the invention.

Textiles

Paisley was the home of a large textile trade, giving the name to the Paisley pattern which originated in the Indian sub-continent but was developed to such a fine art in the West of Scotland. The need for reliable threads is obvious, and it was

The Household

the Paisley firm of J & J Clark who first sold thread on cotton reels instead of in hanks.

Footwear

Jack Sutherland Abbott (1900-79) was born in England at Northampton, but spent many years working with Saxone, the Kilmarnock shoe manufacturers established by his father. Although the company later became part of the British Shoe Company the name remained on the High Streets. Probably better known still is the shoe brand Sutherland launched in 1962, Hush Puppies.

Chemicals

The Edinburgh firm of **MacFarlan Smith** developed Bitrex, the world's most bitter substance, as recognised by the Guinness Book of Records. It is added to household products such as cleaners to deter people from swallowing them.

If Scotland is well known as the home of whisky and haggis, it is also the country of origin of other foods and drinks of almost equal renown. The country's doctors and dentists will also testify to the national sweet tooth.

LIME CORDIAL – A SCOTTISH SUCCESS STORY

- Edinburgh doctor James Lind was the first to recommend the use of lemon and lime juice to combat scurvy

- Glibert Blane, another Scot, introduced limes to naval rations and devised a method of preserving the juice for long journeys

- Lachlan Rose of Leith patented a method of preserving lime juice without alcohol, for the navy had used rum as a preservative. He used sugar to sweeten it, thus devising the first brand of fruit drink. Rose's Lime Juice Cordial, in its distinctive bottle, is more famous now than any other of Lachlan Rose's drinks. Today, though, it is often as a mixer with alcoholic drinks, something its originator may not have approved.

Cola drinks are probably the most popular soft drinks in the world today. The origins of cola or kola drinks is unclear, but the drink was popular in Scotland in the late 19th century, and in 1879 an American writer referred to the drink called Kola. He suggested that an enterprising American manufacturer might consider taking the flavouring, which originated in Africa, and adopting it in the way practised in Scotland. American manufacturers were engaging in an early trade war in the use of coca flavouring, and one, John Pemberton, hit on the idea of combining the flavours of coca and cola in 1884-85 to make the first Coca-Cola.

Food & Drink

Food

James Keiller (1775-1839) of Dundee's entrepreneurial skills helped make the best of a mistake. Having purchased a consignment of Seville oranges which were too bitter for normal consumption, Mrs Janet Keiller used them to make a jelly in the same way she normally used quinces. This was the first batch of many, for in 1797 James Keiller set up a company to produce marmalade in quantity. A further development of marmalade came in Paisley in the 1850s when shreds of orange peel were first added by a grocer to his wife's recipe to create the first Golden Shred. James Robertson's trademark golly is now probably as famous as the product it promotes.

A by-product of the Paisley cotton business became a standard in the kitchen. Starch is a commodity much used in the textile trade, but is also an constituent of many foodstuffs. **John Polson** of Paisley patented various foodstuffs based on corn starch, or cornflour which he first made in 1854, including Brown & Polson's Paisley Flour. The firm also produced Bisto, still the market-leading gravy-thickening.

Abram Lyle (1820-91) was the son of a Greenock shipowner. He followed his father into this line of work, and although his family sold their ships in the 1890s, Lyle Shipping still operates out of Glasgow. Lyle's wider renown lies in the sugar business, a large trade growing out of Glasgow's trade connections with the West Indies. Lyle's connection with sugar began when he was paid with a load of distressed sugar in lieu of freight. He developed a sugar refining business, producing various types of sugar cheaply, but concentrating on one speciality product. That was Lyle's Golden Syrup, a product available to this day, its tin still reflecting the religious nature of the founder with its bible quotation: 'Out of the strong came forth sweetness'. It is said that Lyle never met fellow sugar magnate Sir Henry Tate, and the combined company of Tate & Lyle only formed in 1921.